HOLT

Earth Science
The Physical Setting

NEW YORK

Regents Review Guide
Answer Key
with Test Doctor

HOLT, RINEHART AND WINSTON
A Harcourt Education Company
Orlando • **Austin** • New York • San Diego • London

Cover Photo: John & Eliza Forder/Getty Images

Printed in the United States of America

ISBN-13: 978-0-03-093270-0
ISBN-10: 0-03-093270-X

2 3 4 5 6 7 8 9 10 018 09 08 07

Test Doctor

Earth Structure

REVIEW YOUR UNDERSTANDING

1. (1) Earth is flattened at the poles and bulging at the equator, so the measurements are not the same.
(2) Earth's circumference measured at the poles is slightly smaller.
(3) correct

2. (1) correct
(2) According to the "Inferred Properties of Earth's Interior" diagram in the *Earth Science Reference Tables (ESRT)*, the inner core is solid.
(3) The outer core begins at about 3,000 km. The inner core begins at about 5,000 km.
(4) The inner core has higher pressure.

3. (1) According to the *ESRT*, the density in the outer core ranges from 9.9 to 12.1 g/cm^3.
(2) correct
(3) The outer core is liquid. Density ranges from 9.9–12.1 g/cm^3.
(4) The outer core is liquid.

4. (1) According to the *ESRT*, Earth's core is hotter and denser than the crust.
(2) The core is denser than the crust.
(3) correct
(4) The core is hotter than the crust.

5. (1) According to the *ESRT*, the inner core is solid with an average density of about 13 g/cm^3.
(2) The inner core is solid.
(3) Average density in the inner core is about 13 g/cm^3.
(4) correct

6. (1) The magnetic field does not affect seismic waves.
(2) S-waves will pass through the crust-mantle interface.
(3) correct
(4) S-waves are slower than P-waves. S-waves will not pass through the liquid outer core.

7. (1) correct
(2) P-waves can travel through the outer core.
(3) P-waves can transmit through liquids, S-waves cannot.
(4) P-waves can transmit through liquids, S-waves cannot. Both P-waves and S-waves can transmit through solids.

QUESTIONS FOR REGENTS PRACTICE

1. (1) Oceanic crust is denser because it contains very dense minerals.
(2) correct
(3) Oceanic crust is thinner and denser.
(4) Oceanic crust is thinner.

2. (1) According to the *ESRT*, the pressure in the inner core is greater than 3.0 million atmospheres.
(2) The pressure in the stiffer mantle is less than 2.0 million atmospheres.
(3) correct
(4) The pressure in the plastic mantle is less than 1.0 million atmospheres.

3. (1) correct
(2) According to the *ESRT*, the mantle is denser than the crust and less dense than the outer core.
(3) The mantle is denser than the crust.
(4) The mantle is less dense than the outer core.

4. (1) A volcano is an opening in the crust through which magma and gases escape; the Moho was not found through the study of volcanoes.
(2) This boundary was not discovered by analyzing rocks in the crust.
(3) Scientists can't drill holes deep enough to reach this boundary.
(4) correct

5. (1) S-waves cannot transmit through liquids, only solids.
(2) correct
(3) S-waves are slower than P-waves.
(4) Both P-waves and S-waves travel faster through more-rigid materials.

6. (1) The mantle lies above the outer core.
(2) This presents the order from the uppermost layer inward.
(3) The crust is the outermost layer. The mantle lies above the outer core.
(4) correct

7. (1) According to the *ESRT*, the Moho is the boundary between the crust and the mantle.
(2) The Moho is the boundary between the crust and the mantle.
(3) correct
(4) The Moho is the boundary between the crust and the mantle.

8. (1) According to the *ESRT*, pressure increases steadily with depth. This graph shows density decreasing and then increasing.
(2) This graph shows density increasing with depth, but the density values are not accurate.
(3) correct
(4) This graph shows density decreasing with depth.

Test Doctor

Mapping Earth's Surface

REVIEW YOUR UNDERSTANDING

1. (1) The plane would need to fly south, not north, to reach this location.
(2) 62 – 10 = 52° north of the equator, not 72°.
(3) correct
(4) The plane would need to fly south, not north, to reach this location.

2. (1) According to the *ESRT*, Syracuse is latitude 43°N; Polaris would appear 43° above the horizon, not 90°, which is directly overhead.
(2) This would be the angle above the horizon at the equator.
(3) This would be the angle above the horizon at 23°N.
(4) correct

3. (1) correct
(2) This would happen if the car were traveling north.
(3) Polaris would move closer to the horizon; it would not stay in the same place.
(4) Polaris drops out of sight only at latitudes south of the equator (0°).

4. (1) V-shaped contour lines point upstream, in the direction from which the river flows. The river is not flowing toward the northwest.
(2) correct
(3) The river is flowing toward the northeast.
(4) The river is not flowing from the northwest.

5. (1) correct
(2) Location B has an elevation of about 200 m.
(3) Location C has an elevation of about 100 m.
(4) Location D is in a depression.

6. (1) Index contours are in 100-meter intervals; 100 m ÷ 5 = 20 m.
(2) Each contour line represents 20 meters of elevation.
(3) Each contour line is 20 meters of elevation.
(4) correct

7. (1) correct
(2) Contour lines that are spaced far apart indicate a gradual slope.
(3) Contour lines that are bent to form a V shape are used to indicate the direction from which a river flows.
(4) Contour lines that form closed loops indicate a depression in the land.

QUESTIONS FOR REGENTS PRACTICE

1. (1) The angle of Polaris above the horizon is equal to latitude. 90°N is the North Pole.
(2) Polaris is 0° above the horizon at the equator.
(3) correct
(4) According to the *ESRT*, 35°N is not a latitude in New York State.

2. (1) The change in field value (elevation) is 7,000 feet (8,000 – 1,000); 7,000 feet ÷ 20 miles = 350 ft/mi.
(2) The average gradient is 350 ft/mi.
(3) correct
(4) The average gradient is 350 ft/mi.

3. (1) According to the *ESRT*, Tug Hill Plateau is farther north and west.
(2) The Taconic Mountains are farther north.
(3) The St. Lawrence Lowlands are farther north and west.
(4) correct

4. (1) correct
(2) According to the *ESRT*, these coordinates are too far south and east.
(3) Student confused latitude and longitude. These coordinates are latitude not found in New York State.
(4) These coordinates are too far north and east. The coordinates are not in New York State.

5. (1) The difference between the index contours is 5,000 feet. 5,000 feet ÷ 5 = 1,000 feet.
(2) 5,000 feet ÷ 5 = 1,000 feet.
(3) correct
(4) 5,000 feet ÷ 5 = 1,000 feet.

6. (1) This is the latitude of the equator.
(2) This latitude is halfway between the equator and the North Pole.
(3) correct
(4) This latitude is closer to the South Pole.

7. (1) Contour lines surrounding a river are V-shaped, with the V pointing upstream; lines do not run parallel to the river.
(2) Contour lines do not run perpendicular to the river.
(3) correct
(4) The V shape points upstream, not downstream.

8. (1) correct
(2) Latitude is correct; longitude is not on the grid.
(3) Longitude is correct; latitude is not on the grid.
(4) Coordinates are not on the grid.

9. (1) The angle of Polaris above the horizon equals the latitude of the observer. The latitude of point A is 42.15°N, so Polaris would appear 42.15° above the horizon.
(2) This would be the angle of Polaris above the horizon at 77.8°N.
(3) correct
(4) This would be the angle of Polaris above the horizon at 43.15°N.

10. (1) Coordinates of point A are 42.15°N, 76.8°W. According to the *ESRT*, Buffalo is closer to 43°N, 79°W.
(2) Jamestown is closer to 42°N, 79°W.
(3) Riverhead is closer to 41°N, 73°W.
(4) correct

Test Doctor

Earth Chemistry and Mineral Resources

REVIEW YOUR UNDERSTANDING

1. (1) Color can be observed without changing the substance, so it is a physical property.
(2) correct
(3) Density is a physical property.
(4) Hardness is a physical property.

2. (1) An electron is a subatomic particle.
(2) A proton is a subatomic particle.
(3) The nucleus is the center of an atom.
(4) correct

3. (1) Protons are in the nucleus; electrons are not.
(2) Neutrons are in the nucleus; electrons are not.
(3) correct
(4) Protons and neutrons are in the nucleus; electrons are not.

4. (1) Mass number must be greater than 12, because it includes number of neutrons.
(2) This may be true, but it may be false. Most atoms have unequal numbers of protons and neutrons.
(3) correct
(4) Isotopes are atoms that have the characteristics of the element but different numbers of neutrons.

5. (1) correct
(2) This is not the definition of a mineral. Some minerals can be used as fuel.
(3) Some minerals will sink in water.
(4) Some minerals can be powdered.

6. (1) Diamond (10) can't be scratched.
(2) Biotite mica (2.5–3) can't be scratched by halite (2.5).
(3) correct
(4) Calcite (3) can't be scratched by halite (2.5).

7. (1) Hardness is not related to whether a mineral cleaves.
(2) These minerals are silicates; some silicates cleave, and some do not.
(3) Some minerals that cleave have high density, and some have low density.
(4) correct

8. (1) Volume was divided by mass.
(2) correct
(3) Volume was subtracted from mass.
(4) incorrect placement of decimal point

9. (1) correct
(2) This process forms coal, which is not a mineral ore.
(3) Gasoline is not a mineral ore.
(4) This does not describe the formation of an ore.

10. (1) correct
(2) Sunlight is not in limited supply.
(3) Moving water is not in limited supply.
(4) Wind is not in limited supply.

11. (1) Coal forms from plant remains, not dinosaur fossils.
(2) Coal was not present when Earth was formed.
(3) correct
(4) The original materials have decayed, so there are few recognizable fossils left.

12. (1) correct
(2) Coal takes millions of years to form, so it is nonrenewable.
(3) The amount of gold is limited. New gold cannot be created.
(4) Petroleum takes millions of years to form, so it is nonrenewable.

QUESTIONS FOR REGENTS PRACTICE

1. (1) Mass and volume are needed to determine density.
(2) This cannot be determined without doing physical and chemical tests.
(3) correct
(4) The sample would need to be rubbed on a streak plate.

2. (1) The Mohs scale compares mineral hardness. It has nothing to do with density.
(2) correct
(3) The Mohs scale compares mineral hardness. It has nothing to with mass.
(4) The Mohs scale compares mineral hardness. It has nothing to do with how a mineral cleaves.

3. (1) The color of the streak produced by gypsum is not reported in the "Properties of Common Minerals" table in the *Earth Science Reference Tables (ESRT)*.
(2) The color of the streak produced by calcite is not reported in the *ESRT*.
(3) The color of the streak produced by fluorite is not reported in the *ESRT*.
(4) correct

4. (1) According to the *ESRT*, galena has metallic luster.
(2) correct
(3) According to the *ESRT*, fluorite is harder than dolomite.
(4) Halite doesn't bubble in acid.

5. (1) Mass is not a useful physical property for identification purposes.
(2) According to the *ESRT*, neither mineral contains iron, so a magnet will not help.
(3) correct
(4) According to the *ESRT*, both produce a similar streak.

6. (1) According to the "Properties of Common Minerals" table and the "Scheme for Sedimentary Rock Identification" table in the *ESRT*, pyroxene is a mineral, and limestone is a rock.
(2) According to the *ESRT*, hematite is a mineral, and quartzite is a rock.
(3) According to the *ESRT*, fluorite is a mineral, and coal is a rock.
(4) correct

7. (1) According to the *ESRT*, fluorite cleaves in four directions.
(2) correct
(3) According to the *ESRT*, quartz does not cleave, and it is harder than olivine.
(4) According to the *ESRT*, pyrite does not cleave, it fractures.

8. (1) According to the *ESRT*, color, hardness, and luster can distinguish these minerals.
(2) According to the *ESRT*, color and luster can distinguish these minerals.
(3) According to the *ESRT*, these minerals can be distinguished by color.
(4) correct

9. (1) The order of tests is incorrect.
(2) correct
(3) The order of tests is incorrect.
(4) The order of tests is incorrect.

10. (1) The tests cannot help determine when the minerals formed.
(2) The tests cannot help determine the environment in which the minerals were found.
(3) The tests will not reveal a mineral's chemical properties, only its physical properties.
(4) correct

11. (1) According to the *ESRT*, quartz is hard and has nonmetallic luster.
(2) According to the *ESRT*, pyrite is relatively hard.
(3) correct
(4) According to the *ESRT*, gypsum has nonmetallic luster.

Test Doctor

Rocks

REVIEW YOUR UNDERSTANDING

1. (1) This describes sedimentary rock formation.
(2) correct
(3) This describes igneous rock formation.
(4) This process is called weathering.

2. (1) According to the "Rock Cycle in Earth's Crust" diagram in the *Earth Science Reference Tables (ESRT)*, metamorphic rock forms when heat and/or pressure changes existing rock.
(2) Lava is magma that has erupted onto Earth's surface.
(3) correct
(4) Sediments are solid particles of rock.

3. (1) Erosion is the process by which rock particles are transported from their source to other locations.
(2) Weathering is the process by which rock is broken down into smaller pieces.
(3) correct
(4) This happens to magma during formation of igneous rock.

4. (1) According to the "Scheme for Igneous Rock Formation" in the *ESRT*, granite also contains potassium feldspar and quartz.
(2) correct
(3) Gabbro also contains pyroxene and olivine.
(4) Pegmatite has a very course texture and contains quartz and potassium feldspar.

5. (1) correct
(2) According to the *ESRT*, gabbro is dark-colored and contains no potassium feldspar or quartz.
(3) Basalt is fine-grained, relatively dark-colored, and contains no potassium feldspar or quartz.
(4) Rhyolite is a fine-grained rock.

6. (1) According to the *ESRT*, felsic rocks are light-colored rocks.
(2) Felsic rocks are light-colored rocks.
(3) correct
(4) Felsic rocks have relatively low density.

7. (1) All igneous rocks are formed by cooling magma/lava.
(2) Grain size is largely determined by time, not mineral composition.
(3) Grain size is largely determined by time.
(4) correct

8. (1) According to the "Scheme for Sedimentary Rock Formation" in the *ESRT*, breccia contains angular rock fragments.
(2) Sandstone particles are 0.2 to 0.006 cm in diameter.
(3) Siltstone particles are 0.006 to 0.0004 cm in diameter.
(4) correct

9. (1) According to the *ESRT*, conglomerate contains rounded, pebble-sized rock fragments.
(2) Shale particles are less than 0.0004 cm in diameter.
(3) correct
(4) Sandstone particles are 0.2 to 0.006 cm in diameter.

10. (1) According to the *ESRT*, halite is formed chemically through evaporation.
(2) Conglomerate consists of pebble-sized fragments that have been cemented together.
(3) Gypsum is formed chemically through evaporation.
(4) correct

11. (1) According to the "Scheme for Metamorphic Rock Identification" in the *ESRT*, shale, not breccia, changes to slate during regional metamorphism.
(2) Dolostone does not change to slate during regional metamorphism.
(3) Conglomerate does not change to slate during regional metamorphism.
(4) correct

12. (1) According to the *ESRT*, gneiss has a foliated texture.
(2) correct
(3) Hornfels is fine-grained.
(4) Slate has a foliated texture.

13. (1) According to the *ESRT*, these are both sedimentary rocks; contact metamorphism is related to igneous rocks.
(2) correct
(3) Limestone is a sedimentary rock. Gneiss is a metamorphic rock formed through regional metamorphism.
(4) These are both sedimentary rocks.

14. (1) According to the *ESRT*, quartzite, not schist, forms from the contact metamorphism of quartz sandstone.
(2) Quartzite forms from the contact metamorphism of quartz sandstone.
(3) correct
(4) Quartzite forms from the contact metamorphism of quartz sandstone.

QUESTIONS FOR REGENTS PRACTICE

1. (1) Rocks can contain the same minerals, so this could not be used as a basis for classification.
(2) correct
(3) Rocks can have the same color.
(4) Rocks can be the same age.

2. (1) Extrusive rock forms at Earth's surface.
(2) correct
(3) This choice does not explain why there are no coarse-grained extrusive rocks.
(4) Weathering does not change grain size.

3. (1) According to the *ESRT*, breccia is sedimentary rock.
(2) Sandstone is sedimentary rock.
(3) Gneiss is metamorphic rock.
(4) correct

4. (1) A felsic rock is an igneous rock that contains a large proportion of silica.
(2) correct
(3) Metamorphic rock is categorized by texture (foliated or nonfoliated) and by origin.
(4) A mafic rock is an igneous rock that is rich in iron and magnesium.

5. (1) correct
(2) According to the *ESRT*, felsic rocks contain less iron.
(3) Felsic rocks do not contain olivine.
(4) Felsic rocks do not contain pyroxene.

6. (1) According to the *ESRT*, these processes do not result in formation of sediments.
(2) This process does not result in the formation of sediments.
(3) This process does not result in the formation of sediments.
(4) correct

7. (1) According to the *ESRT*, both granite and basalt are metamorphic rocks.
(2) correct
(3) Granite contains more biotite and amphibole than basalt.
(4) Granite is less dense than basalt.

8. (1) According to the *ESRT*, sand grains are smaller than 2 cm.
(2) correct
(3) Nonclastic rocks are not made of rock fragments.
(4) Nonclastic rocks are not made of rock fragments.

9. (1) According to the *ESRT*, gabbro does not contain quartz.
(2) Granite contains grains from 1 mm to 10 mm in diameter.
(3) correct
(4) Basalt does not contain quartz.

10. (1) According to the *ESRT*, shale changes into slate first; schist should be third.
(2) correct
(3) Shale changes into slate first; gneiss should be last.
(4) Shale changes into slate first; gneiss should be last.

11. (1) correct
(2) According to the *ESRT*, felsic rocks are low in iron and rich in aluminum.
(3) Felsic rocks are low in magnesium and high in aluminum.
(4) Felsic rocks are low in iron and magnesium.

12. (1) According to the *ESRT*, feldspar is found in slate and phyllite.
(2) Mica is found in slate and phyllite.
(3) Quartz is found in slate and phyllite.
(4) correct

Test Doctor

Weathering and Soils

REVIEW YOUR UNDERSTANDING

1. (1) Weathering causes rock to break down physically and chemically.
(2) Weathering causes rock to break down physically and chemically.
(3) Weathering causes rock to break down physically and chemically.
(4) correct

2. (1) This is mechanical weathering.
(2) This is mechanical weathering.
(3) This is mechanical weathering.
(4) correct

3. (1) This is chemical weathering.
(2) correct
(3) This is chemical weathering.
(4) Ice wedging occurs in places that experience alternating freezing and thawing.

4. (1) Burrowing animals do not have a significant effect on the rate of weathering.
(2) correct
(3) Plants do not have a significant effect on the rate of weathering.
(4) If the rock is old, it is most likely weathered already. However, the age of the rock does not alter weathering rate.

5. (1) correct
(2) Roots can split rock into pieces.
(3) The acid dissolves calcite-bearing rocks, such as limestone.
(4) Water expands when it freezes. This expansion can split rocks.

6. (1) correct
(2) According to the "Relationship of Transported Particle Size to Water Velocity" graph in the *Earth Science Reference Tables (ESRT)*, silt is not the smallest particle.
(3) According to the *ESRT*, sand is not the smallest particle.
(4) According to the *ESRT*, pebbles are not the smallest particles.

7. (1) Organic remains are most abundant in the top layer, or A-horizon.
(2) correct
(3) Broken bedrock is found in the third lowest layer, or C-horizon.
(4) Weathering primarily affects the top layer of soil.

8. (1) Grasses and shrubs need water, fertile soil, and time to grow.
(2) correct
(3) Layers of soil must first be weathered and eroded.
(4) The A-horizon would have to be eroded first.

9. (1) Clay particles are the smallest sediment particles. Soils made entirely of clay have low porosity, because there is little space between particles for water to flow.
(2) This is a poorly sorted soil, which has low porosity. Small particles fill the spaces between larger particles, so there is little open space for water to flow.
(3) This is a poorly sorted soil, which has low porosity. Small particles fill the spaces between larger particles, so there is little open space for water to flow.
(4) correct

10. (1) correct
(2) Impermeable means that water cannot flow through it.
(3) Unsaturated means that the soil is not holding much water.
(4) Water cannot infiltrate a soil that is impermeable.

11. (1) As porosity and permeability decrease, infiltration decreases.
(2) correct
(3) As porosity increases, permeability increases as well.
(4) As porosity decreases, permeability decreases as well.

QUESTIONS FOR REGENTS PRACTICE

1. (1) This process does not shape rock.
(2) This process alters the chemical composition of rock, not the shape.
(3) correct
(4) Oxidation is a chemical reaction that changes the chemical composition of rock, not the shape.

2. (1) More surfaces become exposed when a rock breaks apart, so surface area increases.
(2) correct
(3) This is not a result of mechanical weathering.
(4) This is not a result of mechanical weathering.

3. (1) Tightly packed soils have low porosity. There is little open space between particles for water to flow.
(2) correct
(3) In poorly sorted soils, small particles fill the spaces between larger particles, reducing the open space needed for water to flow.
(4) In poorly sorted soils, small particles fill the spaces between larger particles, reducing the open space needed for water to flow.

4. (1) According to the *ESRT*, silt particles are between 0.0004 cm and 0.006 cm in diameter.
(2) Clay particles are less than 0.0004 cm in diameter.
(3) correct
(4) Pebbles are between 0.2 cm and 6.4 cm in diameter.

5. (1) Chemical weathering is the process by which rock is broken down by chemical reactions. This describes a physical process.
(2) Erosion is the process by which rock particles are carried from one location to another.
(3) Infiltration is the process by which water enters and flows through the soil.
(4) correct

6. (1) An increase in rainfall increases the possibility of runoff, because the soil can only hold so much water before it becomes saturated.
(2) An increase in snowmelt increases the possibility of runoff.
(3) correct
(4) An increase in groundwater could mean that the soil is already saturated, which increases the possibility of runoff.

7. (1) correct
(2) This is true, but it tends not to happen in hot, wet climates.
(3) Evaporation of water is not an agent of weathering.
(4) Infiltration of water does not weather rock.

8. (1) correct
(2) In soils that consist of large particles, infiltration rate is high because there is more open space between particles for water to flow. Therefore, the larger the particles, the greater the infiltration rate. This graph says that as infiltration rate increases, particle size decreases.
(3) This graph says that there is no relationship between particle size and infiltration rate.
(4) This graph says that there is a parabolic relationship between particle size and infiltration rate.

9. (1) In unsorted (poorly sorted) soils, small particles fill the spaces between larger particles, reducing the open space needed for water to flow. Therefore, as a soil becomes more unsorted, porosity decreases. This graph says that porosity increases as a soil becomes more unsorted.
(2) correct
(3) This graph says that there is no relationship between particle sorting and porosity.
(4) This graph says that there is a parabolic relationship between particle sorting and porosity.

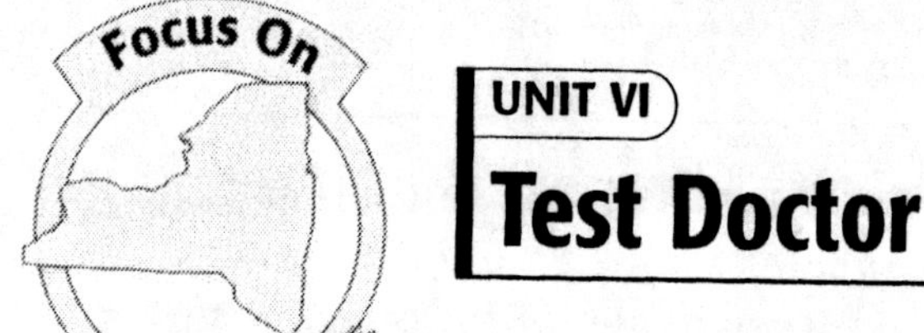

Test Doctor

Erosion and Deposition

REVIEW YOUR UNDERSTANDING

1. (1) correct
(2) According to the "Equations" box in the *Earth Science Reference Tables (ESRT)*, gradient is calculated by dividing the change in field value—in this case 60 m (100 m – 40 m)—by the distance, 100 km. Student incorrectly divided distance by change in field value.
(3) Student added 100 + 40 to get incorrect change in field value.
(4) Student used final field value instead of change in field value.

2. (1) According to the "Relationship of Transported Particle Size to Water Velocity" graph in the *ESRT*, clay particles, which are the smallest particles, would not be deposited until the river slows to less than 20 cm/sec.
(2) Silt particles would not be deposited until the river slows to less than 40 cm/sec.
(3) Sand particles would not be deposited until the river slows to less than 50 cm/sec.
(4) correct

3. (1) According to the *ESRT*, the minimum water velocity needed to carry a boulder is about 300 cm/sec.
(2) This is the minimum velocity needed to carry a small cobble.
(3) This is the minimum velocity needed to carry a large cobble.
(4) correct

4. (1) Color does not affect settling rate.
(2) Low-density particles tend to settle slowly.
(3) Less-rounded, irregularly shaped particles tend to settle more slowly than rounded particles.
(4) correct

5. (1) Wind erosion is significant along the shorelines of large lakes.
(2) Wind erosion is significant in deserts.
(3) correct
(4) Wind erosion is significant in all coastal regions.

6. (1) A ventifact is a sand-blasted rock.
(2) correct
(3) A berm is the raised portion at the back of a beach.
(4) A beach is created by wave action.

7. (1) Dunes are created by wind action, not waves.
(2) A spit is a long, narrow deposit of sand attached at one end to land.
(3) correct
(4) A berm is the raised portion at the back of a beach.

8. (1) A dune is created by wind action.
(2) A barrier island is a well-developed sandbar that forms offshore. Barrier islands are not formed by longshore transport.
(3) A sandbar is an offshore, underwater ridge of sand. It is not formed by longshore transport.
(4) correct

9. (1) Moving ice alone does not create parallel striations.
(2) correct
(3) These processes do not create striations.
(4) The weight of the ice would not create parallel striations.

10. (1) V-shaped valleys are associated with river erosion.
(2) correct
(3) Streams do not necessarily suggest the passage of a glacier.
(4) Boulders do not necessarily suggest the passage of a glacier.

11. (1) Drumlins are not used to determine the age of a glacier.
(2) Drumlins are not used to determine the thickness of a glacier.
(3) correct
(4) Drumlins are not used to determine the rate of glacial movement.

12. (1) This forms a ground moraine.
(2) correct
(3) This forms a lateral moraine.
(4) This forms a medial moraine.

QUESTIONS FOR REGENTS PRACTICE

1. (1) According to the *ESRT*, this is a clay particle, which can be carried in a stream moving about 20 cm/sec.
(2) This is silt, which can be carried in a stream moving about 40 cm/sec.
(3) This is sand, which can be carried in a stream moving about 50 cm/sec.
(4) correct

2. (1) Wind does not create parallel grooves in rock.
(2) Moving water does not create parallel grooves in rock.
(3) correct
(4) Moving water does not create parallel grooves in rock.

3. (1) correct
(2) Density affects the sinking rate of sediments.
(3) Velocity determines the sizes of particles that can be carried by a stream.
(4) Particle size is directly related to deposition.

4. (1) correct
(2) According to the *ESRT*, cobbles would not be deposited until the stream slows to below 300 cm/sec.
(3) Pebbles would not be deposited until the stream slows to below 200 cm/sec.
(4) Sand would not be deposited until the stream slows to below 50 cm/sec.

5. (1) Dynamic equilibrium means that erosion and deposition are in a state of balance.
(2) correct
(3) Dynamic equilibrium means that erosion and deposition are in a state of balance.
(4) Dynamic equilibrium means that erosion and deposition are in a state of balance.

6. (1) Deposition is the release of sediments by an agent of erosion.
(2) Climate influences whether glaciers can exist in an area, but it doesn't control a glacier's movement.
(3) Winds do not cause glacial movement.
(4) correct

7. (1) This is formed by wave deposition.
(2) This is a depositional feature.
(3) correct
(4) This is a depositional feature.

8. (1) Water temperature would not create this pattern.
(2) This would have no effect on sorting.
(3) Particle shape can affect sorting, but it is unlikely that shape alone caused this pattern.
(4) correct

9. (1) correct
(2) According to the *ESRT*, the larger the particle, the greater the water velocity needed to transport that particle. As the water slows, the largest sediment particles are generally deposited first, because they are usually heavier.
(3) High-density materials generally settle quickly.
(4) This is not necessarily true.

Test Doctor

Plate Tectonics

REVIEW YOUR UNDERSTANDING

1. (1) If Antarctica had remained stationary, its climate would not have been able to support tropical vegetation.
(2) This collision would not account for millions of years of vegetation.
(3) correct
(4) Seeds would not germinate in this environment.

2. (1) Tectonic plates diverge, not converge, at mid-ocean ridges.
(2) Rocks at mid-ocean ridges don't age at different rates.
(3) Tectonic forces do not cause rocks to age differently.
(4) correct

3. (1) correct
(2) Plates collide at a convergent boundary.
(3) The plate boundary in California is not a mid-ocean boundary.
(4) Plates move apart at a divergent boundary.

4. (1) According to the "Tectonic Plates" map in the *Earth Science Reference Tables (ESRT)*, this is a divergent boundary. Volcanoes can occur at divergent boundaries, but not due to subduction.
(2) According to the *ESRT*, this is a transform boundary. Volcanic activity would not occur here.
(3) correct
(4) According to the *ESRT*, this is a divergent boundary.

5. (1) Continental crust is less dense, so it would not sink beneath oceanic crust.
(2) correct
(3) This describes a transform boundary. Subduction zones are associated with convergent boundaries.
(4) This describes a divergent boundary.

6. (1) According to the *ESRT*, geologic activity in the Ring of Fire is due to converging plates, not diverging plates.
(2) Geologic activity in the Ring of Fire is due to converging plates.
(3) correct
(4) This doesn't explain the widespread geologic activity in the Ring of Fire.

7. (1) correct
(2) This forms at divergent boundaries.
(3) This could form at a divergent oceanic-oceanic plate boundary or over a mantle hot spot.
(4) This could form as a result of an oceanic-oceanic plate collision.

8. (1) Hot spots occur within plates. Deep-ocean trenches are at the edges of plates.
(2) correct
(3) Subduction zones occur at the edges of plates.
(4) Island arcs occur at the edges of plates.

9. (1) Movement at this boundary can form underwater volcanic mountain ranges.
(2) Movement at this boundary can form folded mountains, not volcanic mountains.
(3) correct
(4) Plates slide past each other at a transform boundary; this movement would not create a volcanic mountain range.

10. (1) Rocks are not faulted when two continental plates collide.
(2) correct
(3) Volcanic mountains are typically associated with oceanic-continental collisions.
(4) Volcanic mountains are typically associated with oceanic-continental collisions.

11. (1) According to the *ESRT*, this coast is a convergent oceanic-continental plate boundary. Mountain building is expected.
(2) According to the *ESRT*, part of this coast is a convergent oceanic-continental boundary. Mountain building is expected.
(3) According to the *ESRT*, this occurs along convergent oceanic-continental boundaries.
(4) correct

12. (1) The epicenter occurs on the surface, directly above the focus.
(2) This describes the total energy of the earthquake.
(3) This describes the damage done by the earthquake.
(4) correct

13. (1) According to the "Earthquake P-wave and S-wave Travel Time" graph in the *ESRT*, the waves would arrive about 2 minutes apart if the earthquake were this distance from the seismograph.
(2) According to the *ESRT*, the waves would arrive less than 4 minutes apart.
(3) correct
(4) According to the *ESRT*, the waves would arrive about 9 minutes apart.

14. (1) According to the *ESRT*, this is a site of diverging plates.
(2) correct
(3) According to the *ESRT*, this area is along a transform boundary.
(4) According to the *ESRT*, this is a site of diverging plates.

15. (1) correct
(2) This is a surface wave, which is the slowest type of seismic waves and the last to arrive at a seismograph station.
(3) This is a surface wave.
(4) S-waves are slower than P-waves.

QUESTIONS FOR REGENTS PRACTICE

1. (1) correct
(2) According to the *ESRT*, this is a convergent boundary.
(3) This is a convergent boundary.
(4) This is a transform boundary.

2. (1) Oceanic crust is denser than continental crust.
(2) correct
(3) Subduction is related to density, not temperature.
(4) Subduction is related to density.

3. (1) Earth's inner core is made of rock and metal and is not related to movement of tectonic plates.
(2) correct
(3) Volcanic eruptions are caused by plate movement, not the reverse.
(4) Earthquakes are caused by plate movement, not the reverse.

4. (1) correct
(2) This would not cause rock layers to bend and fold.
(3) Tension would not cause rocks to fold. Rather, it would create faults (cracks).
(4) Such forces would not fold and bend rock.

5. (1) Both P-waves and S-waves can travel through objects.
(2) correct
(3) P-waves are the fastest seismic waves.
(4) Both P-waves and S-waves can travel through objects.

6. (1) Seismic wave arrival is not related to earthquake strength.
(2) Seismic wave arrival is not related to earthquake strength.
(3) The shorter the lag time, the closer the earthquake.
(4) correct

7. (1) San Francisco is next to the San Andreas Fault. The map indicates that this city has a 21% chance of experiencing a large earthquake between now and 2032.
(2) correct
(3) Unlike the other cities, San Jose is not right next to a major fault line, so it is less likely to experience a major earthquake.
(4) Half Moon Bay has about a 10% chance of experiencing a major earthquake.

8. (1) According to the *ESRT*, this boundary is in Central America.
(2) This boundary is off the coast of South America.
(3) correct
(4) According to the *ESRT*, these Pacific plates do not share a boundary.

9. (1) S-waves are slower than P-waves; thus, they would not be the first to arrive.
(2) S-waves are slower than P-waves. Thus, they would not be the first to arrive.
(3) San Jose is closer to Palo Alto than to San Mateo.
(4) correct

Test Doctor

Earth History

REVIEW YOUR UNDERSTANDING

1. (1) Fossils are also found in disturbed layers.
(2) correct
(3) Tilted layers have been disturbed.
(4) Faulted layers have been disturbed.

2. (1) The skull would not contain measurable amounts of uranium-238.
(2) The skull would not contain measurable amounts of rubidium-87.
(3) correct
(4) The skull would not contain measurable amounts of potassium-40.

3. (1) Weathering does not affect the characteristics of radioisotopes.
(2) Sedimentary rocks usually do not contain the isotopes needed for dating.
(3) correct
(4) Not all radioisotopes are organic in origin.

4. (1) According to the "Radioactive Decay Data" table in the *Earth Science Reference Tables* (*ESRT*), the half-life of potassium-40 is 1.3 billion years. To have a 25% to 75% ratio, the isotope must have passed through two half-lives, or 2.6 billion years.
(2) This number is too low.
(3) correct
(4) This number is too high.

5. (1) This choice does not explain how the jungle animals came to be in a desert-like environment.
(2) It is unlikely that a large population of jungle animals would migrate to a desert-like environment.
(3) correct
(4) This choice does not explain how the jungle animals came to be in a desert-like environment.

6. (1) A trace fossil is a fossil that preserves a trace of an organism's existence. A whole tooth is not a trace; it is part of the organism.
(2) This could be a whole or nearly whole organism.
(3) correct
(4) This is not a mere trace of the organism's existence.

7. (1) According to the "Geologic History of New York State" chart in the *ESRT*, *Centroceras* thrived during the Devonian Period, about 400 million years ago.
(2) This is not the Devonian Period.
(3) correct
(4) This is not the Devonian Period.

8. (1) According to the *ESRT*, *Phacops* lived during the Devonian Period, 362–418 million years ago.
(2) This is not the Devonian Period.
(3) correct
(4) This is not the Devonian Period.

9. (1) According to the *ESRT*, insects appeared before mammal-like reptiles.
(2) Fish appeared before birds.
(3) correct
(4) Coral appeared before extinction of armored fish. Grasses appeared much later.

10. (1) According to the *ESRT*, humans first appeared in the Quaternary Period.
(2) Birds first appeared in the Jurassic Period.
(3) Flowering plants first appeared in the Cretaceous Period.
(4) correct

11. (1) According to the *ESRT*, *Valcouroceras* appeared 443–490 million years ago. *Phacops* appeared much later, so it would not be found in the same age rocks.
(2) *Stylonurus* appeared 362–418 million years ago.
(3) *Bothriolepis* appeared 362–418 million years ago.
(4) correct

12. (1) According to the *ESRT*, Pangaea began breaking up about 250 million years ago, which is in the Mesozoic Era.
(2) Paleozoic Era occurred 251–544 million years ago.
(3) correct
(4) Late Proterozoic is part of the Precambrian Era, which occurred 544 million to 4.6 billion years ago.

QUESTIONS FOR REGENTS PRACTICE

1. (1) To have a 1:3 ratio, the isotope must have passed through two half-lives, or 1.4 billion years.
(2) This number is too low.
(3) correct
(4) This number is too high.

2. (1) By 3.75 million years, the isotope has passed through three half-lives. 16 g ÷ 2 = 8 g (1 half-life); 8 g ÷ 2 = 4 g (2 half-lives); 4 g ÷ 2 = 2 g (3 half-lives). So, 2 g remain.
(2) correct
(3) This is how much would remain after two half-lives.
(4) This is how much would remain after one half-life.

3. (1) Paleozoic ended 251 million years ago, so carbon dating won't work.
(2) correct
(3) Paleozoic ended 251 million years ago, so carbon dating won't work.
(4) Precambrian ended 544 million years ago, so carbon dating won't work.

4. (1) Quaternary Period is the most recent period. Most species appeared earlier.
(2) correct
(3) Fossils are not found in igneous rocks.
(4) Many species appeared before and after the Permian Period.

5. (1) According to the *ESRT*, this event occurred 443–490 million years ago.
(2) This event occurred 142–206 million years ago.
(3) Birds appeared about 150 million years ago.
(4) correct

6. (1) correct
(2) According to the *ESRT*, this happened in the Pleistocene, not the Pennsylvanian.
(3) This happened in the Pleistocene, not the Eocene.
(4) This happened in the Pleistocene, not the Mississippian.

7. (1) correct
(2) According to the *ESRT*, these continents were joined during the Ordovician Period.
(3) These continents were well separated by the Tertiary Period.
(4) These continents were joined during the Silurian Period.

8. (1) This observation would give a relative age, not an absolute age.
(2) correct
(3) This is not a method for dating rocks.
(4) This is not a method for dating rocks.

9. (1) With 5% of the original radioactive isotope remaining, the bone would be much older than this, according to the graph.
(2) According to the graph, by 8,500 years there would be 30–40% remaining.
(3) correct
(4) According to the graph, by 50,000 years there would be less than 5% remaining.

10. (1) Half-life is the amount of time it takes for half (50%) of the parent isotope to decay. The graph shows that it takes about 6,000 years for 50% of the carbon-14 to decay.
(2) correct
(3) By 40,000 years, about 1% of carbon-14 remains.
(4) By 50,000 years, less than 1% of carbon-14 remains.

11. (1) The graph shows that the organic material would have to be close to 20,000 years old if only 10% of carbon-14 remained.
(2) The material would be about 9,000 years old if only 35% of carbon-14 remained.
(3) The material would be about 8,000 years old if only 40% of carbon-14 remained.
(4) correct

Test Doctor

Oceanography

REVIEW YOUR UNDERSTANDING

1. (1) River currents do not cause ocean currents.
(2) correct
(3) Gravity influences all motion, but it is not the cause of ocean currents.
(4) This does not cause ocean currents.

2. (1) The curvature of ocean currents is not related to the moon's rotation.
(2) The curvature of ocean currents is not related to the moon's orbit of Earth.
(3) correct
(4) The curvature of ocean currents is not related to Earth's orbit of the sun.

3. (1) According to the "Surface Ocean Currents" map in the *Earth Science Reference Tables (ESRT)*, the California Current is a cool current.
(2) The Canaries Current is a cool current.
(3) correct
(4) The Peru Current is a cool current.

4. (1) correct
(2) According to the *ESRT*, these currents flow toward the equator.
(3) The Peru and Benguela currents are cool currents.
(4) The Peru and Benguela currents are cool currents that flow toward the equator.

5. (1) Tides are predictable. Tide tables are available months in advance.
(2) Tides are cyclic. Most places have two high tides and two low tides every day.
(3) correct
(4) Tides are cyclic.

6. (1) correct
(2) Spring and neap tides occur every month, regardless of season.
(3) These tides can happen at any time of day during a full or new moon.
(4) These tides can happen on any day.

7. (1) High tides occur at the same time on opposite sides of the world, so they occur about 12 hours apart.
(2) High tides occur about 12 hours apart.
(3) correct
(4) High tides occur about 12 hours apart.

8. (1) More than 70% is covered.
(2) More than 70% is covered.
(3) correct
(4) More than 70% is covered.

9. (1) The continental rise is the gently sloping portion of the continental margin formed by the accumulation of sediments.
(2) correct
(3) A mid-ocean ridge is an underwater mountain range created by diverging plates.
(4) The abyssal plain is the vast, flat part of the deep-ocean basin.

10. (1) correct
(2) The abyssal plain is the vast, flat part of the deep-ocean basin.
(3) The continental rise is the gently sloping portion of the continental margin formed by the accumulation of sediments.
(4) The continental slope is a steep slope on the seaward side of the continental shelf.

11. (1) The continental slope is the part of the continental shelf that slopes steeply.
(2) A submarine valley is a deep canyon on the continental shelf.
(3) An ocean trench is an extremely deep canyon in the deep-ocean basin.
(4) correct

QUESTIONS FOR REGENTS PRACTICE

1. (1) correct
(2) Mid-ocean ridges and trenches are related to movement at plate boundaries.
(3) These features occur in the deep-ocean basin, between continents.
(4) These features are not related to surface currents.

2. (1) According to the *ESRT,* this current flows along the eastern coast of South America.
(2) This current is in the Pacific Ocean, not the Atlantic.
(3) This cool current flows south along the coast of northwestern Africa.
(4) correct

3. (1) According to the *ESRT,* this current flows away from the equator.
(2) correct
(3) This is a warm current.
(4) This current does not flow toward the equator.

4. (1) According to the *ESRT,* the Gulf Stream is in the Atlantic Ocean.
(2) correct
(3) The Gulf Stream is in the Atlantic.
(4) The Gulf Stream is in the North Atlantic.

5. (1) The garbage would be carried by the Gulf Stream, which is in the Atlantic.
(2) correct
(3) The garbage would be carried by the Gulf Stream, which is in the Atlantic.
(4) The garbage would be carried by the Gulf Stream, which is in the Atlantic.

6. (1) High tides, on average, are about 12 hours apart.
(2) High tides, on average, are about 12 hours apart.
(3) High tides, on average, are about 12 hours apart.
(4) correct

7. (1) This describes neap tides.
(2) correct
(3) The total amount of time between tides may differ from day to day, but on average high tides are always about 12 hours apart.
(4) The total amount of time between tides may differ from day to day, but on average high tides are always about 12 hours apart.

8. (1) This is the continental rise.
(2) This is at a seamount, which is an underwater volcanic mountain.
(3) This is a mid-ocean ridge, where plates are diverging.
(4) correct

9. (1) This is the continental rise. It is not a plate boundary.
(2) correct
(3) This is a mid-ocean ridge, where plates diverge.
(4) This is an ocean trench, where plates converge.

10. (1) This is the base of the continental slope. It is not a plate boundary.
(2) This is a trench, which is formed by converging plates.
(3) correct
(4) This is the continental slope. It is not a plate boundary.

Test Doctor

The Atmosphere

REVIEW YOUR UNDERSTANDING

1. (1) In the upper troposphere, solar radiation is not scattered by clouds and water vapor; therefore radiation is more intense.
(2) correct
(3) The density of the air does not cause the temperature decrease in the troposphere.
(4) This would not cause temperature to decrease.

2. (1) According to the "Selected Properties of Earth's Atmosphere" graph in the *Earth Science Reference Tables (ESRT)*, pressure this low is not observed in troposphere.
(2) correct
(3) Pressure in mesosphere measures below 10^{-3} atm.
(4) Thermosphere has no measurable atmospheric pressure.

3. (1) According to the *ESRT*, atmospheric pressure this low is found 50–80 km above sea level.
(2) Pressure this low is found 50–80 km above sea level.
(3) correct
(4) This measurement is off the scale.

4. (1) correct
(2) According to the *ESRT*, the stratosphere has no measurable water vapor content.
(3) The mesosphere has no measurable water vapor content.
(4) The thermosphere has no measurable water vapor content.

5. (1) correct
(2) Most reflected sunlight escapes into space.
(3) Convection moves heat away from the lowest portion of the atmosphere to help warm other areas.
(4) Scattering does not transfer heat energy.

6. (1) According to the "Electromagnetic Spectrum" chart in the *ESRT*, wavelengths of 10^{-5} cm are ultraviolet waves. The troposphere does not contain sufficient ozone to absorb UV waves.
(2) correct
(3) The mesosphere does not contain sufficient ozone to absorb UV waves.
(4) The thermosphere does not contain ozone.

7. (1) According to the *ESRT*, these are ultraviolet waves, not infrared.
(2) This is red light.
(3) correct
(4) These are microwaves and radio waves.

8. (1) correct
(2) When air is heated, air molecules spread apart, so the air becomes less dense.
(3) When air is heated, air molecules spread apart, so the air becomes less dense.

9. (1) The rotation of Earth affects wind direction, but it does not create wind.
(2) Humidity differences are not the cause of wind.
(3) correct
(4) The revolution of Earth does not cause wind.

10. (1) Gravitational forces do not curve winds right or left.
(2) correct
(3) The revolution of Earth is not related to the Coriolis effect.
(4) Incoming solar radiation does not curve winds.

11. (1) Polar air creates areas of high pressure.
(2) correct
(3) Easterly winds blow from east to west.
(4) Polar air creates areas of high pressure. Easterly winds blow from east to west.

QUESTIONS FOR REGENTS PRACTICE

1. (1) Convection is the transfer of heat by movement of fluids, due to density differences.
(2) When light is reflected, the energy is not transferred to Earth.
(3) correct
(4) Conduction is the transfer of heat through direct contact. The sun and Earth are not in contact.

2. (1) correct
(2) According to the *ESRT,* air temperature increases as altitude decreases in the mesosphere.
(3) In the mesosphere, air temperature increases as altitude decreases.
(4) In the mesosphere, air temperature increases as altitude decreases.

3. (1) correct
(2) According to the *ESRT,* there is no measurable water vapor in stratosphere. Atmospheric pressure is less than 10^{-1} atmospheres.
(3) There is no measurable water vapor in stratosphere. Atmospheric pressure is less than 10^{-1} atmospheres.
(4) There is no measurable water vapor in stratosphere. Atmospheric pressure is less than 10^{-1} atmospheres.

4. (1) correct
(2) According to the "Planetary Wind and Moisture Belts in the Troposphere" diagram in the *ESRT,* these winds are the westerlies, which come from the southwest in the Northern Hemisphere.
(3) These winds are the westerlies, which come from the southwest in the Northern Hemisphere.
(4) These winds are the westerlies, which come from the southwest in the Northern Hemisphere.

5. (1) According to the *ESRT,* polar easterlies are north of the United States.
(2) Northeast trades are south of the United States, near the equator.
(3) correct
(4) Northwest trades are south of the equator.

6. (1) Air near equator is warm, not cool.
(2) Warm air has low density.
(3) Air near equator is warm, not cool.
(4) correct

7. (1) Smooth-textured materials tend to reflect sunlight, not absorb it.
(2) correct
(3) Light-colored and smooth-textured materials tend to reflect sunlight.
(4) Light-colored materials tend to reflect sunlight.

8. (1) Winds curve to the left in the Southern Hemisphere.
(2) correct
(3) Winds are deflected to the right in the Northern Hemisphere and to the left in the Southern Hemisphere.
(4) Winds are deflected to the right in the Northern Hemisphere.

9. (1) correct
(2) The magnitude of the Coriolis effect increases moving away from equator.
(3) The magnitude of the Coriolis effect increases moving away from equator.
(4) There is no Coriolis effect at the equator.

Test Doctor

Weather

REVIEW YOUR UNDERSTANDING

1. (1) According to the "Relative Humidity" chart in the *Earth Science Reference Tables (ESRT)*, the relative humidity is 45%. To find relative humidity, look for the dry-bulb temperature in the far-left column of the chart. The difference between the dry-bulb and wet-bulb temperatures is 6°C, so move your finger across the row to the right until you find the "6" column. The number in the box is the relative humidity.
(2) The relative humidity is 45%.
(3) The relative humidity is 45%.
(4) correct

2. (1) According to the "Dewpoint Temperatures" chart in the *ESRT*, the dew point is 1°C. To find dew point, look for the dry-bulb temperature in the far-left column of the chart. The difference between the dry-bulb and wet-bulb temperatures is 2°C, so move your finger across the row to the right until you find the "2" column. The number in the box is the dew point temperature.
(2) The dew point is 1°C.
(3) correct
(4) The dew point is 1°C.

3. (1) Dew point is the temperature to which air must be cooled in order to reach the saturation point—100% relative humidity.
(2) This means that the air holds only 25% of the water it could hold at that temperature.
(3) This means that the air holds only 50% of the water it could hold at that temperature.
(4) correct

4. (1) According to the "Pressure" scale in the *ESRT*, 1004.0 millibars is the same as 29.65 inches of mercury.
(2) correct
(3) An air pressure of 1004.0 millibars is equivalent to 29.65 inches of mercury.
(4) An air pressure of 1004.0 millibars is equivalent to 29.65 inches of mercury.

5. (1) Winds are not directly caused by differences in air temperature.
(2) Winds are not directly caused by differences in air temperature.
(3) correct
(4) Winds blow from high-pressure areas to low-pressure areas, not the reverse.

6. (1) Evaporation is the change in matter from a liquid to a gaseous state. Air does not evaporate. Liquid water can evaporate.
(2) Air blows toward the center of a low.
(3) correct
(4) The inflow of air does not create a high because air is rising from the center.

7. (1) At a warm front, warm air rises and cools, producing clouds and precipitation.
   (2) Warm air rises.
   (3) correct
   (4) Air does not get warmer as it rises. Temperature decreases with altitude.

8. (1) Temperature decreases when a cold front arrives.
   (2) Temperature decreases when a cold front arrives. Cold fronts also produce clouds and precipitation.
   (3) correct
   (4) Cold fronts produce clouds and precipitation.

9. (1) In the Northern Hemisphere, air circulates counterclockwise around a low.
   (2) Air circulates counterclockwise and toward the center of a low.
   (3) correct
   (4) Air circulates toward the center of a low.

10. (1) Station models summarize weather conditions on the ground near a weather station.
    (2) correct
    (3) Station models summarize weather conditions near a weather station.
    (4) Station models summarize weather conditions near a weather station.

11. (1) correct
    (2) The line points to the direction from which winds are moving.
    (3) The line points to the direction from which winds are moving.
    (4) The line points to the direction from which winds are moving.

12. (1) Isobars are lines that connect points of equal pressure, not temperature.
    (2) Isobars indicate current air pressure, not changes in air pressure.
    (3) correct
    (4) Isobars do not indicate moving fronts.

QUESTIONS FOR REGENTS PRACTICE

1. (1) Have students consult page 243 for information on how to get relative humidity from psychrometer readings.
   (2) The relative humidity is 40%.
   (3) correct
   (4) The relative humidity is 40%.

2. (1) Air reaches the saturation point when air temperature equals dew point. These temperature readings indicate a dew point of 14°C. Therefore, the air would have to cool 6°C. See page 243 for information on how to find dew point.
   (2) correct
   (3) The air would have to cool 6°C.
   (4) The air would have to cool 6°C.

3. (1) correct
   (2) This is the same as 1004.0 millibars.
   (3) Pressure scale was read incorrectly.
   (4) Pressure scale was read incorrectly.

4. (1) The question describes a warm front, but this answer describes a cold front.
   (2) correct
   (3) Air temperature increases and pressure decreases with the arrival of a warm front.
   (4) Air temperature increases with the arrival of a warm front.

5. (1) Air moves clockwise and away from the center of a high in the Northern Hemisphere.
   (2) Air moves clockwise around a high in the Northern Hemisphere.
   (3) Air moves away from the center of a high.
   (4) correct

6. (1) correct
   (2) According to the "Weather Map Symbols" in the *ESRT*, this represents a warm front.
   (3) This represents an occluded front.
   (4) This represents a stationary front.

7. (1) This is 10 knots, or 11.5 mi/h. The solution requires that the wind speed be converted to knots. According to the "Weather Map Symbols" in the *ESRT*, 1 knot is equal to 1.15 mi/h. Therefore, 52 mi/hr is equal to about 45 knots. One whole feather on a station model indicates a wind speed of 10 knots. A half feather is 5 knots. Thus, 4.5 feathers would indicate a wind speed of 52 mi/h.
(2) This is 15 knots, or 17.25 mi/h.
(3) This is 30 knots, or 34.5 mi/h.
(4) correct

8. (1) Relative humidity is 50% and air temperature is 25°C.
(2) Graph was read incorrectly.
(3) Graph was read incorrectly.
(4) correct

9. (1) Temperature increased about 15 Celsius degrees over the six-hour period. Therefore, the rate of increase is about 2.5° per hour.
(2) correct
(3) The rate of increase is 2.5° per hour.
(4) The rate of increase is 2.5° per hour.

10. (1) At 2:00 A.M. evaporation would be slow, because air temperature is relatively low.
(2) At 8:00 A.M., evaporation would be slow because air temperature is relatively low.
(3) correct
(4) At 6:00 P.M., air temperature is relatively high, but not as high as it was at 4:00 P.M.

Test Doctor

Climate

REVIEW YOUR UNDERSTANDING

1. (1) This is near the Antarctic Circle. Here, insolation strikes Earth at a low angle, so temperatures are cold.
(2) correct
(3) This is the North Pole.
(4) This is a middle-latitude region. Temperatures are warm, but not warmer than areas closer to the equator.

2. (1) Land and water receive the same amount and intensity of insolation.
(2) Water actually reflects more insolation.
(3) Water has a lower density.
(4) correct

3. (1) correct
(2) Specific heat of land does not change with increasing distance from the equator.
(3) Insolation strikes Earth at a lower angle in northern latitudes than at the equator.
(4) Even if the humidity were greater, it would not affect temperature.

4. (1) Greenhouse effect occurs because greenhouse gases absorb and trap heat.
(2) correct
(3) This does not explain why they are called greenhouse gases.
(4) This does not explain why they are called greenhouse gases.

5. (1) Burning of fossil fuels does not cause ice ages.
(2) correct
(3) Burning of fossil fuels does not increase the frequency of El Niño events.
(4) Burning of fossil fuels does not cause volcanic eruptions.

6. (1) Global warming is a gradual increase in global temperature.
(2) An ice age is a period of very low global temperatures.
(3) correct
(4) The greenhouse effect is the natural process by which gases in the atmosphere absorb and trap heat.

7. (1) Rainfall increases during an El Niño, because warm air rises over the warming ocean, creating clouds and precipitation.
(2) correct
(3) Air pressure drops and rainfall increases during an El Niño.
(4) Air pressure drops during an El Niño.

QUESTIONS FOR REGENTS PRACTICE

1. (1) Land absorbs heat rapidly.
(2) Land does not have a high specific heat.
(3) Ocean water is not a good conductor of heat; it tends to release heat slowly.
(4) correct

2. (1) This is close to the South Pole, where insolation intensity is very low.
(2) This is the South Pole, where insolation intensity is very low.
(3) correct
(4) This is a middle-latitude zone, where insolation intensity is not as high as in equatorial locations.

3. (1) correct
(2) Very little precipitation occurs at the poles.
(3) Temperature is low at the poles.
(4) Temperature is low, and little precipitation occurs here.

4. (1) correct
(2) Carbon dioxide is a greenhouse gas. Greenhouse gases absorb heat and trap it close to Earth's surface.
(3) Absorption of ultraviolet radiation does not contribute to global warming.
(4) Absorption of ultraviolet radiation does not contribute to global warming.

5. (1) According to the "Planetary Wind and Moisture Belts" diagram in the *Earth Science Reference Tables (ESRT)*, air pressure is low and humidity high at 0°, which is the equator.
(2) This location is in a tropical zone, where air pressure is low and humidity high.
(3) correct
(4) This location is in a polar zone, where pressure and humidity are high.

6. (1) This fact is not related to the physical properties of water.
(2) Density does not influence water's ability to hold heat energy.
(3) correct
(4) This fact does not influence water's ability to hold heat energy.

7. (1) Air becomes warm and dry as it moves downslope.
(2) Air becomes warm and dry as it moves downslope.
(3) Air becomes cooler as it moves upslope. This would increase its relative humidity but not the absolute amount of moisture it contains.
(4) correct

8. (1) Ultraviolet radiation does not influence the temperature of an area.
(2) correct
(3) The equator may have more cloud cover than the North Pole, but this is not the reason for the temperature differences.
(4) The thickness of the atmosphere is the same at the equator as it is at the North Pole.

9. (1) A higher elevation would make City B cooler all year.
(2) This would not affect the temperature.
(3) They are at the same latitude, so they should receive the same amount of insolation.
(4) correct

10. (1) Tropical climates have much warmer temperatures.
(2) correct
(3) Polar climates have much colder temperatures.

Test Doctor

Movements of Earth and Its Moon

REVIEW YOUR UNDERSTANDING

1. (1) correct
(2) Earth's rotation is related to the day.
(3) The sun does not revolve around Earth.
(4) The sun's rotation is not related to Earth's year.

2. (1) Earth's revolution is related to the year.
(2) correct
(3) The sun does not revolve around Earth.
(4) The sun's rotation is not related to Earth's day.

3. (1) The shape of Earth's orbit does not influence the movement of a Foucault pendulum.
(2) Earth's revolution around the sun does not influence the movement of a Foucault pendulum.
(3) correct
(4) The rate at which Earth revolves around the sun does not influence the movement of a Foucault pendulum.

4. (1) The month is roughly based on this relationship.
(2) correct
(3) Standard units of time are not based on this relationship.
(4) Standard units of time are not based on this relationship.

5. (1) This represents the summer season.
(2) This represents the fall season.
(3) correct
(4) This represents the spring season.

6. (1) Distance from the sun is not the cause of seasons.
(2) Distance from the sun is not the cause of seasons.
(3) correct
(4) The sun gives off the same amount of energy over the course of the year.

7. (1) This marks the start of the spring season. The sun is not at its highest point in the sky.
(2) correct
(3) This marks the start of the winter season. The sun is not at its highest point in the sky.
(4) This marks the start of the fall season. The sun is not at its highest point in the sky.

8. (1) The moon has less mass than Earth, so its gravity is weaker than Earth's gravity.
(2) If this were true, the moon would not rotate.
(3) correct
(4) Earth does not revolve and rotate at the same rate.

9. (1) A solar eclipse could occur at this position.
(2) No eclipse would occur at this position.
(3) correct
(4) No eclipse would occur at this position.

10. (1) The side of the moon facing Earth is unlit during the new moon phase.
(2) correct
(3) One-half of the side of the moon facing Earth is visible at first quarter.
(4) One-half of the side of the moon facing Earth is visible at last quarter.

QUESTIONS FOR REGENTS PRACTICE

1. (1) This is the definition of a year.
(2) This is 27.3 days, a sidereal month.
(3) correct
(4) This is 27.3 days; the same as the moon's period of revolution; a sidereal month.

2. (1) Duration and intensity decrease from 5/21 to 12/21.
(2) correct
(3) Intensity and duration increase together and decrease together.
(4) Intensity and duration increase together and decrease together.

3. (1) It takes about 29.5 days for a complete cycle of phases.
(2) It takes about 29.5 days for a complete cycle of phases.
(3) correct
(4) It takes about 29.5 days for a complete cycle of phases.

4. (1) This does happen every month.
(2) This describes a lunar eclipse.
(3) This never happens.
(4) correct

5. (1) The sun does rotate, but this has nothing to do with its apparent path in the sky.
(2) The sun does not revolve around Earth.
(3) correct
(4) This has no effect on the sun's apparent path in the sky.

6. (1) correct
(2) This would occur whether Earth was tilted or not.
(3) This is not related to Earth's tilt. It is an effect of distance.
(4) This is related to Earth's rotation, not tilt.

7. (1) This has no effect on the noontime height of the sun.
(2) correct
(3) This has no effect on the noontime height of the sun.
(4) This has no effect on the noontime height of the sun.

8. (1) The sun would be higher in the sky on this date.
(2) The sun would be at its highest point in the sky for the year on this date.
(3) correct
(4) On this date, the sun would be slightly higher than it is at position A.

9. (1) correct
(2) This does not happen at this latitude.
(3) This happens at the winter solstice.
(4) This does not happen at this latitude.

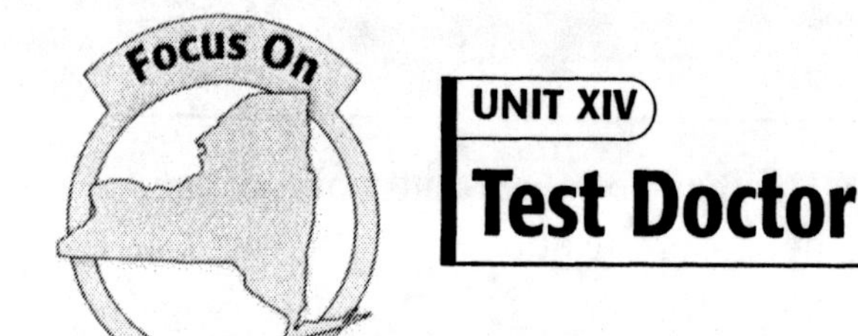

Test Doctor

The Solar System

REVIEW YOUR UNDERSTANDING

1. (1) correct
(2) Mars does not orbit Earth in the heliocentric model.
(3) The sun does not orbit Earth in the heliocentric model.
(4) Polaris does not orbit Earth in the heliocentric model.

2. (1) According to the "Equations" in the *Earth Science Reference Tables (ESRT)*, eccentricity equals the distance between foci divided by the length of the major axis. There are no units of measure for eccentricity.
(2) There are no units of measure for eccentricity.
(3) correct
(4) Eccentricity is always less than 1. Student divided major axis by distance between foci.

3. (1) Orbital speed is not related to rotational speed.
(2) correct
(3) Eccentricity describes the orbit as a whole; it does not vary from point to point in the orbit.
(4) Tilt is not related to orbital speed.

4. (1) correct
(2) Eccentricity is always less than 1. Student divided major axis by distance between foci.
(3) There are no units of measure for eccentricity.
(4) Eccentricity is always less than 1. Student divided major axis by distance between foci. There are no units of measure for eccentricity.

5. (1) Student should consult the "Solar System Data" table in the *ESRT* for correct order.
(2) Student should refer to the *ESRT*.
(3) Student should refer to the *ESRT*.
(4) correct

6. (1) According to the *ESRT*, Jupiter is about five times farther ($778.3 \div 149.6$).
(2) correct
(3) Uranus is about 19 times farther.
(4) Mars is about one and one-half times farther.

7. (1) Meteors are meteoroids that enter Earth's atmosphere.
(2) correct
(3) Comets are chunks of ice and dust that orbit the sun in long, elliptical paths.
(4) Moons are bodies that orbit planets.

8. (1) correct
(2) Earth is not larger than Jupiter.
(3) Earth is more solid and denser than Jupiter.
(4) Earth is more solid, denser, and smaller than Jupiter.

9. (1) A protoplanet is an accumulation of rock, ice, and dust that will eventually form a planet.
(2) A planetisimal is a small accumulation of rock, ice, and dust that will eventually join with other planetisimals to form protoplanets.
(3) A star is a large ball of gases that gives off light and heat.
(4) correct

10. (1) correct
(2) About 99.9% of the mass of the solar system is concentrated in the sun.
(3) About 99.9% of the mass of the solar system is concentrated in the sun.
(4) About 99.9% of the mass of the solar system is concentrated in the sun.

11. (1) This describes an early theory of the origin of the solar system, which is not supported by available evidence.
(2) This describes the formation of the sun.
(3) Protoplanets formed in the outer parts of the disk, not the center.
(4) correct

12. (1) All the planets formed from protoplanets.
(2) All the planets originated in the solar nebula.
(3) correct
(4) Only Jupiter, Saturn, Uranus, and Neptune have these characteristics.

QUESTIONS FOR REGENTS PRACTICE

1. (1) correct
(2) Mercury originated from one of the four inner protoplanets.
(3) Venus originated from one of the four inner protoplanets.
(4) Mars originated from one of the four inner protoplanets.

2. (1) Geocentric means "Earth-centered." Heliocentric means "sun-centered."
(2) correct
(3) Both models are based on observations of the sky.
(4) Both models are based on observations of the sky.

3. (1) Jovian planets are larger. Refer student to *ESRT.*
(2) correct
(3) Refer student to *ESRT.*
(4) Refer student to *ESRT.*

4. (1) According to the *ESRT,* Mars is about half the size of Earth.
(2) Saturn is larger than Uranus.
(3) correct
(4) Mercury is about three-quarters the size of Mars.

5. (1) According to the *ESRT,* Mars has a slightly longer period of rotation and is less dense.
(2) Mars is less dense than Earth.
(3) Mars has a slightly longer period of rotation.
(4) correct

6. (1) Gravity is related to mass. However, even if Pluto had the same mass as Mercury, the gravitational attraction between Pluto and the sun would be weaker than the gravitational attraction between Mercury and the sun, because Pluto is very far away from the sun.
(2) correct
(3) The difference in orbital speed between these two planets is primarily related to distance from the sun.
(4) The difference in orbital speed between these two planets is primarily related to distance from the sun.

7. (1) According to the *ESRT,* the eccentricity of Mercury's orbit is 0.206 and the eccentricity of the moon's orbit is 0.055. Mercury's orbit is shaped more like a circle than the moon's orbit.
(2) Neptune's orbit is much more elliptical in shape than the moon's orbit.
(3) Jupiter's orbit is similar in shape to the moon's orbit, but not as similar as Saturn's orbit.
(4) correct

8. (1) The closer the eccentricity is to zero, the closer the orbit is to a perfect circle. According to the *ESRT*, the eccentricity of Pluto's orbit is 0.250, which means that Pluto has more of an oval-shaped orbit than Venus, whose orbit has an eccentricity of 0.007.
(2) Mercury's orbit has an eccentricity of 0.206.
(3) Earth's orbit has an eccentricity of 0.017.
(4) correct

9. (1) The closer the exoplanet is to the star, the greater the gravitational attraction. Gravity will not remain the same because the orbit is not a perfect circle.
(2) Gravity will decrease as the exoplanet moves farther from the star, but will increase again when it moves closer again in the orbit, after point D.
(3) Gravity will not increase as the exoplanet moves away from the star.
(4) correct

10. (1) Orbital speed is greatest when the exoplanet is closest to the star.
(2) correct
(3) Orbital speed is greatest when the exoplanet is closest to the star.
(4) At this point, the exoplanet would be moving at the slowest speed.

Test Doctor

Stars, Galaxies, and the Universe

REVIEW YOUR UNDERSTANDING

1. (1) According to the "Luminosity and Temperature of Stars" graph in the *Earth Science Reference Tables (ESRT)*, the sun has a surface temperature of about 5,500°C. Rigel has a surface temperature of about 11,000°C.
(2) Sirius has a surface temperature of about 10,000°C.
(3) correct
(4) Barnard's star has a surface temperature of about 3,000°C.

2. (1) According to the *ESRT*, this star is hotter and dimmer than the sun.
(2) correct
(3) This star has about the same brightness and temperature as the sun.
(4) This star is cooler and dimmer than the sun.

3. (1) According to the *ESRT*, the sun is a main-sequence star. Blue supergiants are located in the upper left of the diagram.
(2) correct
(3) The sun has a surface temperature of about 5,500°C.
(4) On the diagram, white dwarfs are located below the main sequence.

4. (1) Red giants are cool and luminous.
(2) Blue supergiants are hot and luminous.
(3) Main-sequence stars have a wide range of magnitudes and temperatures, but those that are hot are luminous and those that are cool are dim.
(4) correct

5. (1) correct
(2) Stars like the sun create energy through nuclear fusion of hydrogen into helium.
(3) Stars like the sun create energy through nuclear fusion of hydrogen into helium.
(4) Stars like the sun create energy through nuclear fusion of hydrogen into helium.

6. (1) The sun is a main-sequence star. The main-sequence stage precedes the white dwarf stage.
(2) correct
(3) The sun is a main-sequence star. The sun is not massive enough to become a blue supergiant.
(4) The sun is not massive enough to become a supergiant. Supergiant stars do not become dwarf stars.

7. (1) correct
(2) The sun will likely end its life cycle as a white dwarf.
(3) Procyon B is a white dwarf, so it is at the end of its life cycle.
(4) Barnard's Star is a red dwarf, so it is at the end of its life cycle.

8. (1) Stars in the main sequence exhibit a variety of colors. Color is related to surface temperature.
(2) Stars in the main sequence exhibit a range of surface temperatures.
(3) Stars in the main sequence exhibit a range of luminosities.
(4) correct

9. (1) The universe formed about 13 billion years ago.
(2) The universe formed about 13 billion years ago.
(3) correct
(4) The universe formed about 13 billion years ago.

10. (1) A red shift means that objects are moving away from us, not toward us.
(2) correct
(3) A red shift means that objects are moving away from us, which means that the universe is expanding.

11. (1) Most galaxies are moving away from us, and the more distant galaxies are moving faster.
(2) Most galaxies are moving away from us, and the more distant galaxies are moving faster.
(3) The more distant galaxies are moving faster.
(4) correct

QUESTIONS FOR REGENTS PRACTICE

1. (1) According to the "Luminosity and Temperature of Stars" graph in the *ESRT*, blue stars have surface temperatures above 30,000°C.
(2) correct
(3) Red stars have surface temperatures below 3,500°C.
(4) White stars have surface temperatures between 7,500°C and 10,000°C.

2. (1) A shift in the absorption lines of a star's spectrum indicates that the star is moving.
(2) A shift in the absorption lines of a star's spectrum indicates that the star is moving.
(3) correct
(4) A red shift indicates that the star is moving away.

3. (1) According to the *ESRT*, the sun is dimmer than Betelgeuse.
(2) Procyon B is dimmer than Betelgeuse.
(3) Alpha Centauri is dimmer than Betelgeuse.
(4) correct

4. (1) correct
(2) Blue stars are the hottest stars, with surface temperatures of more than 30,000°C.
(3) Yellow stars are relatively cool, but they are not the coolest stars.
(4) White stars have surface temperatures between 7,500°C and 10,000°C.

5. (1) Blue main-sequence stars are 10,000 to 1,000,000 times brighter than the sun.
(2) The sun is a yellow main-sequence star.
(3) correct
(4) Red main-sequence stars are dimmer than the sun.

6. (1) Distance is not plotted on an H-R diagram.
(2) Color is equivalent to temperature, so they only locate a star on the x-axis of the H-R diagram.
(3) Size and distance are not plotted on the H-R diagram.
(4) correct

7. (1) correct
(2) In a blue shift, wavelengths are shorter than normal, because the object is moving toward us.
(3) In a red shift, wavelengths are longer than normal, because the object is moving away from us.
(4) A blue shift indicates that the object is moving toward us, not away from us.

8. (1) A red shift means the galaxies are moving away from us. A red shift tells nothing about the age of the universe.
(2) correct
(3) A red shift tells nothing about the size of the universe.
(4) A red shift is not used to determine the distribution of matter in the universe.

9. (1) White dwarfs are in region A.
(2) correct
(3) Supergiants are in region D.
(4) Red giants are in region C.

10. (1) The stars in region A are white dwarfs, which are smaller and hotter than the sun. The smallest stars are located near the bottom of the graph.
(2) correct
(3) The stars in region A are white dwarfs, which are smaller and hotter than the sun.
(4) The stars in region A are white dwarfs, which are smaller and hotter than the sun.

11. (1) The stars in region C are giants. The stars in region D are supergiants. In general, supergiants are hotter than giants and exhibit a wider range of colors.
(2) correct
(3) Giants are dimmer than supergiants. In general, supergiants are hotter than giants and exhibit a wider range of colors.
(4) Giants are dimmer than supergiants.

Answer Key

Unit I Earth Structure

REVIEW YOUR UNDERSTANDING

1. (3)
2. (1)
3. (2)
4. (3)
5. (4)
6. (3)
7. (1)

QUESTIONS FOR REGENTS PRACTICE

1. (2)
2. (3)
3. (1)
4. (4)
5. (2)
6. (4)
7. (3)
8. (3)

9. The S-waves were not able to pass through Earth's outer core.
10. outer core
11. The outer core is liquid. P-waves can travel through liquids, solids, and gases, but S-waves can travel only through solids.
12. This layer is called the stiffer mantle.
13. Between points A and B, temperature would increase from about 2,500°C to about 5,000°C. The pressure would increase from less than 1 atmosphere to about 1.5 atmospheres.
14. Location C is in the outer core. Temperature is about 5,500°C. Pressure is about 2.5 atmospheres, and the state of matter is liquid. Location D is in the inner core. Temperature is more than 6,000°C. Pressure is about 3.5 atmospheres, and the state of matter is solid.

Unit II Mapping Earth's Surface

REVIEW YOUR UNDERSTANDING

1. (3)
2. (4)
3. (1)
4. (2)
5. (1)
6. (4)
7. (1)

QUESTIONS FOR REGENTS PRACTICE

1. (3)
2. (3)
3. (4)
4. (1)
5. (3)
6. (3)
7. (3)
8. (1)
9. (3)
10. (4)

11. Contour interval is 20 feet; 100 feet ÷ 5 = 20 feet.
12. Elevation of location C is slightly more than 320 feet.
13.

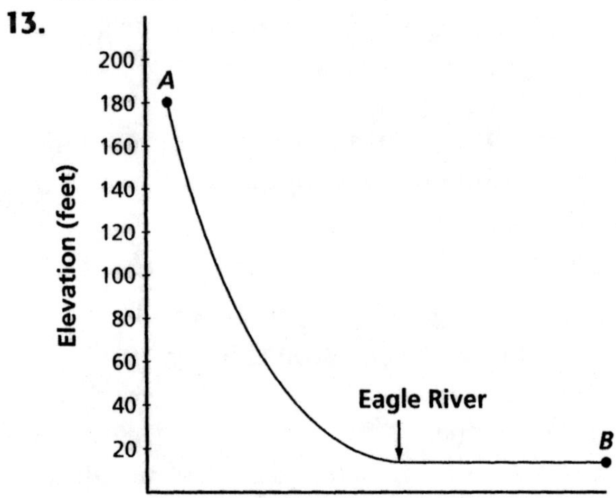

14. V-shaped contour lines can indicate the direction that the river flows. The V points in the direction from which the river flows, so the Cohocton River is flowing northwest to southeast.
15. According to the "Generalized Landscape Regions of New York State" map and the "Generalized Bedrock Geology of New York State" map in the *Earth Science Reference Tables,* the mapped area is part of the Allegheny Plateau.
16. The change in elevation from point X to point Y is 100 ft. The distance between these points is approximately 40 km. Therefore, gradient = 100 ft ÷ 40 km = 2.5 ft/km.

Unit III Earth Chemistry and Mineral Resources

REVIEW YOUR UNDERSTANDING

1. (2)
2. (4)
3. (3)
4. (3)
5. (1)
6. (3)
7. (4)
8. (2)
9. (1)
10. (1)
11. (3)
12. (1)

QUESTIONS FOR REGENTS PRACTICE

1. (3)
2. (2)
3. (4)
4. (2)
5. (3)
6. (4)
7. (2)
8. (4)
9. (2)
10. (4)
11. (3)
12. Streak is more useful for identifying a mineral sample than color is, because many minerals have the same color. For example, biotite mica has a black to dark brown color but produces a white streak.
13. This has to do with hardness. Magnetite has a hardness of 5.5–6.5, and quartz has a hardness of 7. Therefore, quartz can scratch magnetite because it is harder. Magnetite can't be scratched by halite, because halite has a hardness of 2.5. Halite is softer than magnetite.
14. Acceptable answers: I would do a hardness test on the two samples. Quartz has a hardness of 7, and halite has a hardness of 2.5. So, quartz is harder than halite; I would look at the cleavage patterns of the samples. Their cleavage is different. Halite cleaves into cubes when it breaks. Quartz does not cleave, it fractures.
15. The density of a fragment is 2.7 g/cm^3. The density of a fragment is the same as the density of the whole sample. Density is a physical property.
16. The mass is 5.4 grams.
17. The density is 1.66 g/cm^3.
18. Minerals A and C could be the same. Although the colors of these two minerals are different, many minerals come in different colors. Minerals A and C have the same density and the same streak color.
19. The mineral could be galena. Galena has metallic luster, exhibits cleavage, leaves a gray-black streak, and has a density of 7.6 g/cm^3.

Unit IV Rocks

REVIEW YOUR UNDERSTANDING

1. (2)
2. (3)
3. (3)
4. (2)
5. (1)
6. (3)
7. (4)
8. (4)
9. (3)
10. (4)
11. (4)
12. (2)
13. (2)
14. (3)

QUESTIONS FOR REGENTS PRACTICE

1. (2)
2. (2)
3. (4)
4. (2)
5. (1)
6. (4)
7. (2)
8. (2)
9. (3)
10. (2)
11. (1)
12. (4)
13. Rock A is an extrusive igneous rock.
14. Rock A is coarse-grained and contains percentages of olivine, plagioclase feldspar, biotite, and amphibole minerals.
15. Rock C has the largest mineral grains because it is an intrusive igneous rock that formed by slow cooling of magma underground.
16. Rock B is limestone. It could have formed when calcite precipitated out of water. It could also be formed from the shells of marine animals, such as clams and oysters.
17. Rock A is quartz sandstone; Rock B is quartzite; Rock C could be granite or pegmatite.
18. Heat and/or pressure
19. Natural agents such as moving water, wind, gravity, or glaciers could have carried the quartz particles and deposited them in another location.

Answer Key

Unit V Weathering and Soils

REVIEW YOUR UNDERSTANDING

1. (4)
2. (4)
3. (2)
4. (2)
5. (1)
6. (1)
7. (2)
8. (2)
9. (4)
10. (1)
11. (2)

QUESTIONS FOR REGENTS PRACTICE

1. (3)
2. (2)
3. (2)
4. (3)
5. (4)
6. (3)
7. (1)
8. (1)
9. (2)

10. Possible answers: Abrasion due to wind produces rounded and smooth surfaces. Plant roots create cracks in rock. Ice wedging expands cracks. Carbonic acid in rainwater dissolves limestone.

11. Answers may vary. If the rock is moved to an especially cold or dry climate, ice wedging will not occur. Mechanical and chemical weathering by plant life will cease. Limestone will not dissolve. If the rock is moved to a warmer, wetter climate, weathering (particularly chemical weathering) is likely to occur at an increased rate.

12. Answers may vary. Gravity pulls soil particles downhill. Surface runoff rushes down the slope, carrying soil particles with it. Wind carries soil particles off the slope.

13. Possible answers: Pour gravel on hillsides to increase porosity and permeability, which will increase rate of infiltration and decrease runoff. Pour topsoil over clay to increase porosity and permeability, which will increase rate of infiltration and decrease runoff. Plant vegetation to anchor soil and reduce erosion due to gravity, wind, and runoff. Reduce slope of the hill, which will reduce erosion caused by gravity and runoff.

Unit VI Erosion and Deposition

REVIEW YOUR UNDERSTANDING

1. (1)
2. (4)
3. (4)
4. (4)
5. (3)
6. (2)
7. (3)
8. (4)
9. (2)
10. (2)
11. (3)
12. (2)

QUESTIONS FOR REGENTS PRACTICE

1. (4)
2. (3)
3. (1)
4. (1)
5. (2)
6. (4)
7. (3)
8. (4)
9. (1)

10. A

11. The average gradient is 0.5 m/km.

12. The fast-moving water on the outside of the curve rapidly erodes the river bank, enlarging the curve. At the same time, deposition occurs on the inside of the curve, where water is moving more slowly.

13. The waves were moving toward the northwest.

14. barrier island

15. A barrier island begins as a sandbar, which is created when waves carry sand from the beach and deposit it offshore. When the sandbar accumulates enough material to break the water surface, it is called a barrier island. The barrier island will continue to grow in size as more sand is deposited on it and as organic matter accumulates.

Unit VII Plate Tectonics

REVIEW YOUR UNDERSTANDING

1. (3)
2. (4)
3. (1)
4. (3)
5. (2)
6. (3)
7. (1)
8. (2)
9. (3)
10. (2)
11. (4)
12. (4)
13. (3)
14. (2)
15. (1)

QUESTIONS FOR REGENTS PRACTICE

1. (1)
2. (2)
3. (2)
4. (1)
5. (2)
6. (4)
7. (2)
8. (3)
9. (4)
10. 11:01 A.M.
11. The distance between the earthquake epicenter and this seismograph station is approximately 2.8×10^3 kilometers.
12. 60 seconds = 1 minute
    If P-wave travels at 8 km/sec, then it can travel 480 km/min (change to minutes)
    Distance from epicenter to seismograph is 2,800 km
    $$\frac{2{,}800 \text{ km}}{480 \text{ km/min}} = 5.8 \text{ minutes}$$
    Travel time of P-waves was 5.8 minutes, so earthquake occurred at approximately 10:55 A.M.
13. two
14. Pacific plate, North American plate, Eurasian plate, Indian-Australian plate, Nazca plate, Cocos plate
15. It is a convergent oceanic-continental plate boundary.
16. The Ring of Fire occurs along the convergent plate boundaries of several plates. The Pacific plate is subducting beneath the other major tectonic plates, forming subduction zones. As the subducting plate is driven into the mantle, it heats up and begins to melt. Magma erupts onto the surface, forming volcanoes along the plate boundaries.

Unit VIII Earth History

REVIEW YOUR UNDERSTANDING

1. (2)
2. (3)
3. (3)
4. (3)
5. (3)
6. (3)
7. (3)
8. (3)
9. (3)
10. (4)
11. (4)
12. (3)

QUESTIONS FOR REGENTS PRACTICE

1. (3)
2. (2)
3. (2)
4. (2)
5. (4)
6. (1)
7. (1)
8. (2)
9. (3)
10. (2)
11. (4)
12. brachiopods
13. Silurian Period
14. *Eospirifer* is an index fossil. Scientists know that *Eospirifer* lived 418–443 million years ago. Therefore, the rock layer in which an *Eospirifer* fossil is found must be of similar age.
15. Scientists would measure the radioactive materials in the rock in which the fossil was found. They would compare the amount of original radioisotope—for example, uranium-238—with the amount of its decay product, lead-206, in order to determine how many half-lives have elapsed since the rock formed. The number of elapsed half-lives would be multiplied by the half-life of uranium-238 to determine the absolute age of the rock.
16. Fossil F is oldest.
17. Fossil F must be the oldest, because the law of superposition states that a sedimentary rock layer is older than the layers above it.
18. unconformity
19. Acceptable answers: *Cryptolithus, Valcouroceras, Lichenaria, Maclurites*

Unit IX Oceanography

REVIEW YOUR UNDERSTANDING

1. (2)
2. (3)
3. (3)
4. (1)
5. (3)
6. (1)
7. (3)
8. (3)
9. (2)
10. (1)
11. (4)

QUESTIONS FOR REGENTS PRACTICE

1. (1)
2. (4)
3. (2)
4. (2)
5. (2)
6. (4)
7. (2)
8. (4)
9. (2)
10. (3)

11. A and C
12. This picture shows the relative positions of the sun, moon, and Earth during a neap tide. During a neap tide, high tides are lower than usual and low tides are higher than usual, creating a smaller tidal range.
13. The tidal range was greatest on 10/14/04. The tidal range was 3.7 ft.
14. Neap tides occurred on 10/6, 10/21, and 11/5.
15. This was a neap tide. During a neap tide, the sun's gravitational pull counteracts the moon's gravitational pull. As a result, high tides are lower than usual and low tides are higher than usual.

Unit X Atmosphere

REVIEW YOUR UNDERSTANDING

1. (2)
2. (2)
3. (3)
4. (1)
5. (1)
6. (2)
7. (3)
8. (1)
9. (3)
10. (2)
11. (2)

QUESTIONS FOR REGENTS PRACTICE

1. (3)
2. (1)
3. (1)
4. (1)
5. (3)
6. (4)
7. (2)
8. (2)
9. (1)

10. from innermost layer to outermost layer: troposphere (tropopause), stratosphere (stratopause), mesosphere (mesopause)
11. stratosphere
12. The weather balloon will experience a steady decrease in air pressure from point B to point A.
13. carbon dioxide, water vapor; troposphere

14.

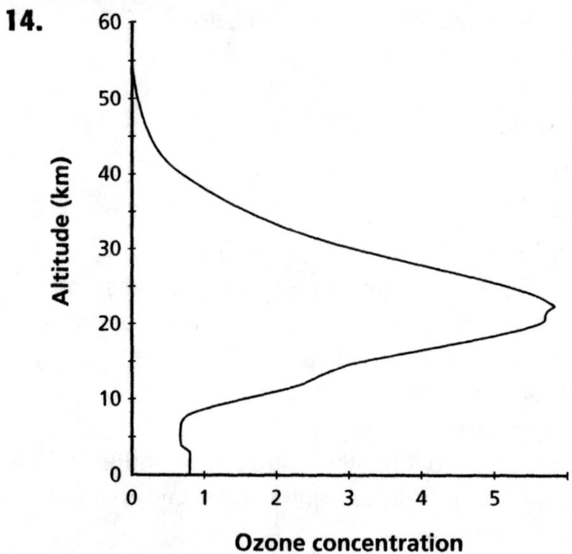

15. stratosphere
16. Ozone absorbs incoming ultraviolet waves.

Unit XI Weather

REVIEW YOUR UNDERSTANDING

1. (4)
2. (3)
3. (4)
4. (2)
5. (3)
6. (3)
7. (3)
8. (3)
9. (3)
10. (2)
11. (1)
12. (3)

QUESTIONS FOR REGENTS PRACTICE

1. (3)
2. (1)
3. (4)
4. (2)
5. (4)
6. (1)
7. (4)
8. (4)
9. (2)
10. (3)

11. Check diagrams for correct use and placement of weather symbols, dew point, air temperature, and air pressure.
12. Acceptable answer: A cold front moved in, causing the warm air to rise into the atmosphere. Water vapor in the air condensed, producing clouds and snow. The drop in air pressure produced a pressure gradient, which resulted in faster winds.
13. There are four fronts: cold front, warm front, occluded front, and stationary front.

14. The air mass over Alabama is a maritime tropical air mass, which is warm and moist. A continental polar air mass, which is cold and dry, covers most of the rest of the United States.
15. A cold front is moving in on Alabama.
16. The cold front will cause the warm, moist air to rise and cool in the atmosphere. Water vapor will condense, producing clouds and precipitation.

Unit XII Climate

REVIEW YOUR UNDERSTANDING

1. (2)
2. (4)
3. (1)
4. (4)
5. (2)
6. (2)
7. (3)
8. (2)

QUESTIONS FOR REGENTS PRACTICE

1. (4)
2. (3)
3. (1)
4. (1)
5. (3)
6. (3)
7. (4)
8. (2)
9. (4)
10. (2)

11. The air cools and the moisture in it condenses, creating precipitation.
12. At location A, the air is warm and moist. At B, the air is warm and dry.
13. The temperature dropped by 0.4°C or 0.5°C.
14. The eruption released ash, dust, and sulfur particles into the atmosphere. This blocked sunlight from reaching Earth's surface, thereby reducing worldwide temperatures.

Unit XIII Movements of Earth and Its Moon

REVIEW YOUR UNDERSTANDING

1. (1)
2. (2)
3. (3)
4. (2)
5. (3)
6. (3)
7. (2)
8. (3)
9. (3)
10. (2)

QUESTIONS FOR REGENTS PRACTICE

1. (3)
2. (2)
3. (3)
4. (4)
5. (3)
6. (1)
7. (2)
8. (3)
9. (1)

10. The shadow is longer on December 21 because the sun is lower in the sky than it is on both September 23 and March 21, which are the autumnal and vernal equinoxes, respectively. On these dates, the shadows are the same length because the noontime sun is at the same height.
11. Because the sun is higher in the sky on June 21 than at any other time of the year, the drawing should show a shadow that is shorter than the shadows in diagrams A and C.
12. Acceptable answers: Tropic of Cancer; 23.5° north latitude; A
13. Location A would have more daylight hours, because the sun would appear higher in the sky.
14. Diagram should show the North Pole tilted away from the sun.

Unit XIV Solar System

REVIEW YOUR UNDERSTANDING

1. (1)
2. (3)
3. (2)
4. (1)
5. (4)
6. (2)
7. (2)
8. (1)
9. (4)
10. (1)
11. (4)
12. (3)

QUESTIONS FOR REGENTS PRACTICE

1. (1)
2. (2)
3. (2)
4. (3)
5. (4)
6. (2)
7. (4)
8. (4)
9. (4)
10. (2)

11. Gravitational attraction increases as the exoplanet moves closer to the star; it decreases the farther the exoplanet is from the star.

12. The mass of GJ876c is 0.6 the mass of Jupiter, which makes it about twice the mass of Saturn.
13. Exoplanet GJ876b is closer to star Gliese 876 than Mercury is to the sun. Therefore, the period of revolution would be somewhat less than 88 days.
14. Eccentricity = 0.27
15. Eccentricity of Earth's orbit is 0.017. Earth's orbit is more circular in shape than the orbit of GJ876c.
16. Gravitational force is strongest at point A, where GJ876c is closest to the star. Gravitational force is least at point B, where GJ876c is farthest from the star.

Unit XV Stars, Galaxies, and the Universe

REVIEW YOUR UNDERSTANDING

1. (3)
2. (2)
3. (2)
4. (4)
5. (1)
6. (2)
7. (1)
8. (4)
9. (3)
10. (2)
11. (4)

QUESTIONS FOR REGENTS PRACTICE

1. (2)
2. (3)
3. (4)
4. (1)
5. (3)
6. (4)
7. (1)
8. (2)
9. (2)
10. (2)
11. (2)
12. Galaxy A shows a red shift, so it is moving away from the observer.
13. Galaxy B shows a blue shift, so it is moving toward the observer.
14. Betelgeuse has the greatest absolute magnitude, so it has the greatest overall energy output.
15. Absolute magnitude is a measure of a star's actual brightness. Apparent magnitude is the brightness of a star as seen from Earth. The sun has a large apparent magnitude because, compared to other stars, it is extremely close to Earth—only 0.000002 light-years.
16. Answers may vary. Possible answer: Of stars in the table, Betelgeuse has the greatest absolute magnitude; it is the brightest star. However, Betelgeuse has a relatively low apparent magnitude. The star appears relatively dim, because it is very far from Earth.

Unit XVI

Page 217

1. The mineral sample has a hardness of 2.5–3.0.
2. The mineral sample has a hardness of 5.5–6.0.

Page 218

1. The density is 2.8 g/cm^3.
2. The volume is 2.1 cm^3.
3. The mass is 5.3 g.

Page 219

1. The relative humidity is 21%.
2. The relative humidity is 89%.